Frank Williams
562-547-4564

Model Program for Special Inspection

Based on Chapter 17 of the 2012 IBC®

A guideline to assist building officials in the administration and enforcement of special inspection code provisions of the 2012 *International Building Code*®

INTERNATIONAL ACCREDITATION SERVICE
Leading Accreditors Since 1975

ICC
INTERNATIONAL CODE COUNCIL®

Model Program for Special Inspection
Based on Chapter 17 of the 2012 IBC

ISBN: 978-1-60983-199-8

The information contained in this document is believed to be accurate; however, it is being provided for informational purposes only and is intended for use only as a guide. Publication of this document by ICC should not be construed as ICC engaging in or rendering engineering, legal or other professional services. Use of the information contained in this publication should not be considered by the user as a substitute for the advice of a registered professional engineer, attorney or other professional. If such advice is required, you should seek the services of a registered professional engineer, licensed attorney or other professional.

Publication Date: March 2012

Printed in the U.S.A.

38385-T015919 4/14

FOREWORD

This document provides the building official with guidance on the administration and implementation of the special inspection requirements of Chapter 17 of the 2012 *International Building Code*® (IBC®). The guidance is based on recommended best practices, and the consensus of building officials, design professionals, and inspection and testing agencies. Duties and responsibilities of the building official, special inspector, project owner, engineer or architect of record, and contractor are defined in this guide.

In the past, special inspection tasks were often delegated to the project owner, architect or engineer of record, and sometimes building contractors. In the 21st century, the responsibilities of the building official in providing oversight of special inspection activities are driven by several factors, including legal liability, insistence on competent performance and completion of work as specified and approved, and more involvement of local government. Accordingly, the use of certified individuals or accredited special inspection agencies is becoming increasingly prevalent and necessary. Section 110.4 authorizes the building official to accept reports of approved inspection agencies provided such agencies have been found to satisfy the requirements as to qualifications and reliability.

Under the IBC, special inspection is not a discretionary activity, nor does it replace required structural observations as required by the design professional in responsible charge. The conditions under which special inspection must be utilized are clearly stated in Section 1704 of the IBC; however, there is a provision for the building official to waive special inspection for minor work.

A special inspector is an individual with specialized skills who observes critical building or structural features identified by the design professional responsible for compliance with plans and specifications approved by the building official. Ensuring the competence of special inspectors or special inspection agencies is the responsibility of the building official. Although the IBC lacks specific qualification requirements, this fact does not lessen the importance of diligence in the critical process of approving special inspectors.

Recognizing the need to support its member building departments across the United States and globally, the International Code Council® (ICC®) developed certification and accreditation programs to identify competent individuals and organizations that perform special inspection. ICC Certification and Testing programs require individuals seeking certification as special inspectors to have demonstrated a level of competence in a manner that will protect the safety and welfare of the public. Individuals may become certified through ICC by participating in a number of seminars and training classes, and passing written examinations. Organizations that employ special inspectors and provide inspection services to the building industry can gain accreditation through International Accreditation Service™ (IAS™), a subsidiary of ICC.

The IAS accreditation program requires special inspection organizations to operate under a documented quality management system, to use certified inspectors and to meet comprehensive accreditation requirements, including on-site assessment by subject-matter experts who are either currently practicing or supervising special inspectors or who are building officials. Successful on-site assessment leads to accreditation for a specific scope of special inspection where the organization has been found competent.

Further information regarding the certification of individuals or accreditation of organizations that perform special inspection can be found on the following websites:

IAS   www.iasonline.org
ICC   www.iccsafe.org/certification

About the International Code Council®

The International Code Council (ICC®), a membership association dedicated to building safety, fire prevention and energy efficiency, develops the codes and standards used to construct residential and commercial buildings, including homes and schools. The mission of ICC is to provide the highest quality codes, standards, products and services for all concerned with the safety and performance of the built environment. Most United States cities, counties and states choose the International Codes, building safety codes developed by the International Code Council. The International Codes also serve as the basis for construction of federal properties around the world, and as a reference for many nations outside the United States. The Code Council is also dedicated to innovation and sustainability and Code Council subsidiary, ICC Evaluation Service, issues Evaluation Reports for innovative products and reports of Sustainable Attributes Verification and Evaluation (SAVE).

Headquarters: 500 New Jersey Avenue, NW, 6th Floor, Washington, DC 20001
District Offices: Birmingham, AL; Chicago, IL; Los Angeles, CA
1-888-422-7233
www.iccsafe.org

ACKNOWLEDGEMENTS

The *Model Program for Special Inspection* debuted in 1986 when it was first offered to building officials, jurisdictions, and inspection and testing agencies. It focused on the field inspection aspects of special inspection and is now used as a standard reference for the special inspector exams.

The original program was the culmination of the efforts of building officials, inspectors and inspection/testing agencies associated with California Council of Testing and Inspection Agencies (CCTIA). Since that time, many others have voluntarily given their time and talents to keep this document current and pertinent to the special inspection profession. This document has been reviewed over the years by special inspector exam committees before and after the consolidation of ICC. The following were participants on these committees:

Charles Ballou	Minnesota	R.D. Messer	Louisiana
Stan Beeler	Ohio	Jim Messersmith	Virginia
Jeffery Bowers	Washington	Charles Mlodzik	Washington
James Byrnes	Kansas	William Murchinson	North Carolina
Clifford Craig	California	Robert Nagin	Florida
Michelle Craig	California	Phillip Nishikawa	California
Fred Deis	Oregon	Sam Palmer	Nevada
David Drenth	Minnesota	Gary Peterson	Washington
Carl Eriksson	Utah	Anthony Re	Oregon
Roger Evans	Utah	Don Riley	Alabama
Fayez Fanik	New Jersey	Gerald Smith	Alabama
William Fitzjohn	California	Bill Stevens	South Carolina
Harry Gleich	South Carolina	Thomas Stevens	California
Charles Hemsley	California	Emile Troup	Maine
Jeff Hilfiker	Kansas	Zan Turner	California
Merlyn Isaak	California	W.F. Urbick	Washington
David Janifer	Virgina	George Voigt	Pennsylvania
Don Johnson	Florida	Michael Wheeler	California
Joe Kane	New York	Kip Williamson	California
Roy Keck	Georgia	David Wismer	Pennsylvania
Ron Lynn	Nevada	Daniel Zechmeister	Michigan
Mehrdad Maher	California		
Calvin McCall	North Carolina		

Thanks to all who have participated in bringing together a model program for special inspection that can be adapted by others without changing the intent of the performance of the special inspector's work.

The accreditation information that was added in this edition was provided by Pat McCullen, Vice President/Chief Technical Officer of IAS.

William Fitzjohn has provided invaluable assistance in reviewing this document through many editions.

(This page is intentionally left blank.)

Contents

(This page is intentionally left blank.)

I. SPECIAL INSPECTION – AN OVERVIEW

There are several areas of construction regulated by the *International Building Code*® (IBC®) where special inspection is mandatory. These inspections are to verify that work that is considered critical to life safety and property protection is being or has been constructed according to the approved plans and specifications. These special inspections are in addition to the typical municipal inspections required by the building department that are specified in Section 110 or specific structural observation as may be required by Section 1704.5. Section 1704 details when special inspection is required and for what types of work. The registered design professional in responsible charge is required to prepare a "Statement of Special Inspections" that outlines the special inspections to be performed.

Special inspection is the monitoring of materials, installation, fabrication, erection and placement of components and connections that require special expertise that are critical to the integrity of the building structure. Special inspection is required to ensure compliance with the approved construction documents (plans) and standards referenced in the applicable codes. Special inspectors or special inspection agencies must be approved by the building official to conduct certain types of inspections as detailed in Section 1704. Although Section 1704.1 requires the owner or the registered design professional acting as the owner's agent to provide for specially qualified inspectors, the approval of special inspectors and special inspection agencies is solely the responsibility of the building official.

Special inspection requirements in the 2012 IBC are comprehensive and consist of 16 major categories as follows:

1. **Inspection of fabricators.** Where fabrication of structural load-bearing members and assemblies is being performed on the premises of the fabricator.

   Note the exception for approved fabricators (see Section 1704.2.5).

2. **Steel construction.** Steel elements of buildings and structures requiring special inspection are found in Section 1705.2. See Table 1705.2.2 for detailed information regarding inspections and referenced standards.

   Specific areas listed in Table 1705.2.2 for steel construction other than structural steel are as follows:

   - Material verification of cold-formed steel deck.
   - Inspection of welding for cold-formed steel deck and reinforcing steel.

   Note the exception in Section 1705.2 discussing steel fabrication without heating operations of any kind.

3. **Concrete construction.** The special inspection and verification requirements for concrete construction are found in Section 1705.3. See Table 1705.3 for detailed information regarding inspections.

   Specific areas listed in Section 1705.3 for concrete construction:

   - Reinforcing steel, including prestressing tendons and placement.
   - Reinforcing steel welding.
   - Anchors cast in concrete.
   - Anchors post-installed in hardened concrete members.
   - Verifying use of required design mix.
   - Sampling fresh concrete, and performing slump, air content and fresh concrete temperature at time of making specimens for strength tests.

- Proper application techniques for concrete and shotcrete placement.
- Maintaining specified curing temperature and techniques.
- Prestressed concrete, including application of prestressing forces and grouting of bonded pre-stressing tendons.
- Erecting precast concrete members.
- Verifying in-situ concrete strength prior to stressing of tendons in post-tensioned concrete, and prior to the removal of shores and forms from beams and structural slabs.
- Formwork for shape, location and dimensions of concrete members.

4. **Masonry construction.** The special inspection and verification requirements for masonry construction are found in Sections 1705.4 and 1705.4.2. Masonry construction shall be inspected and verified in accordance with TMS 402/ACI 530/ASCE 5 and TMS 602/ACI 530.1/ASCE 6 quality assurance program requirements.

5. **Wood construction.** Inspection of the fabrication of wood structural elements and assemblies for high-load diaphragms, both prefabricated and field assembled or site built, are found in Section 1705.5. Special inspections of the fabrication process of prefabricated wood structural elements and assemblies shall be in accordance with Section 1704.2.5. Special inspections of site-built assemblies shall be in accordance with Section 1705.5.

   Note the exception for approved fabricators in Section 1704.2.5.1.

6. **Soils.** Inspection of site for existing conditions, verification of site preparation prior to placement of prepared fill, verification of fill material and maximum lift thicknesses, and verification that in-place densities meet the requirements of the approved soils report (see Section 1705.6 and Table 1705.6).

   Note the exception in this section pertaining to moisture content.

7. **Driven deep, cast-in-place and helical pile foundations.** Inspections performed during the installation and testing of pile foundations (see Tables 1705.7 and 1705.8; and Sections 1705.7, 1705.8 and 1705.9).

8. **Special inspection for wind resistance.** Special inspections itemized in Sections 1705.10.1 through 1705.10.3 are required for buildings and structures constructed in certain wind exposure categories.

   a. **Structural wood.** Continuous inspection of field gluing operations of the main windforce-resisting system is required. Periodic special inspection is required for nailing, bolting, anchoring and other fastening of components (see Section 1705.10.1).

      Note the exception for components of the main windforce-resisting system where fastener spacing of the sheathing is greater than 4 inches (102 mm) on center.

   b. **Cold-formed steel light-frame construction.** Periodic inspection is required during welding operations of elements, and for screw attachment, bolting, anchoring and other fastening of components within the main windforce-resisting system (see Section 1705.10.2 and note the two exceptions as applicable).

9. **Special inspection for seismic resistance.** Section 1705.11 requires inspections listed in Sections 1705.11.1 through 1705.11.8, unless exempted by the exceptions of Section 1704.2.

   a. **Designated seismic systems.** Section 1705.11.4 requires the special inspector to examine designated seismic systems that require qualification in accordance with Section 1705.12.3 to verify that the anchorage or mounting conforms to the certificate of compliance.

b. **Architectural components.** Periodic special inspection is required during the erection and fastening of exterior cladding, interior and exterior nonbearing walls, and interior and exterior veneer in structures assigned to Seismic Design Category D, E or F (see Section 1705.11.5 and note the three exceptions).

c. **Access floors.** Section 1705.11.5.1 requires periodic special inspection for the anchorage of access floors in structures assigned to Seismic Design Category D, E or F.

d. **Mechanical and electrical components.** Section 1705.11.6 contains requirements for periodic special inspection for mechanical and/or electrical equipment for installations in various seismic design categories.

e. **Storage racks.** Periodic special inspection is required for the anchorage of storage racks 8 feet (2438 mm) or greater in height in structures assigned to Seismic Design Category D, E or F (see Section 1705.11.7).

f. **Seismic isolation systems.** Section 1705.11.8 requires special inspection during the fabrication and installation of isolator units and energy dissipation devices for seismic isolation systems.

10. **Testing and qualification for seismic resistance.** Testing and qualification, as specified in Sections 1705.12.1 through 1705.12.4, is required, unless exempted from special inspection by the exceptions of Section 1704.2.

a. **Concrete reinforcement.** The section contains special requirements for reinforcement complying with ASTM A 615 used to resist earthquake-induced flexural and axial forces. Section 1705.12.1 requires compliance with specific sections of ACI 318 and weldability tests to comply with Section 3.5.2 of ACI 318.

b. **Structural steel.** Testing of structural steel shall be conducted in accordance with the quality assurance requirements of AISC 341. Section 1705.12.2 notes an exception for structural steel in structures assigned to Seismic Design Category C.

c. **Seismic certification of nonstructural components.** Section 1705.12.3 requires the registered design professional to specify certifications of nonstructural components and designated seismic systems for compliance with Section 13.2 of ASCE 7 where certification is required by Section 1705.12.

11. **Sprayed fire-resistant materials.** Special inspection of fire-resistant material applied to structural elements and decks in accordance with Sections 1705.13.1 through 1705.13.6. These sections refer to specific elements and components of construction, and provide minimum requirements for in-situ testing.

Inspection of sprayed fire-resistant materials is as follows:

- Verification of structural member surface conditions.
- Verification of the application of materials in accordance with the manufacturer's instructions.
- Verification of the thicknesses and density of applied materials.
- Verification of the bond strength of applied materials.

12. **Mastic and intumescent fire-resistant coatings.** Inspections for mastic and intumescent fire-resistant coatings for structural elements and decks (see Section 1705.14).

13. **Exterior insulation and finish systems (EIFS).** See Section 1705.15 for exceptions to inspection when EIFS is applied over water-resistive barriers with a means for draining excess water, and EIFS installed on masonry or concrete.

14. **Fire-resistant penetrations and joints.** Section 1705.16 covers special inspections for throug penetrations, membrane-penetration firestops, fire-resistant joint systems and perimeter fire barrier systems. These materials are required to be tested and listed in accordance with Sections 714.3.1.2, 714.4.1.2, 715.3 and 715.4 (see Sections 1705.16.1 and 1705.16.2 for additional special inspection requirements).

15. **Special inspection for smoke control.** Section 1705.17 requires the testing of smoke control systems by a special inspector. Special inspection agencies must have expertise in fire protection engineering, mechanical engineering and certification as air balancers (see Section 1705.17.2).

16. **Special cases.** Inspections, in the opinion of the building official, that are needed because of the use of alternative materials, unusual design or use of materials not having building code approval or needing to meet special manufacturer requirements. See Section 1705.1.1 for examples where the building official has discretion to require special inspection.

    *IAS offers accreditation of special inspection agencies for special inspections.

II. GENERAL PROGRAM GUIDELINES

A. Purpose of Special Inspection

Special inspection is the monitoring of the materials and workmanship that are critical to the integrity of the building structure. It is the review of the work of the contractors and their employees to ensure that the approved plans and specifications are being followed, and that relevant codes and ordinances are being observed. The special inspection process is in addition to those conducted by the municipal building inspector and by the design professional in responsible charge as part of periodic structural observation. Special inspectors furnish continuous or periodic inspection as prescribed in Tables 1705.2.2, 1705.3, 1705.6, 1705.7 and 1705.8 for that construction that requires their presence (see Sections 110 and 1704).

Good communication between the special inspector and the designers, contractor and building department is essential to project quality assurance. The following sections, B through F, are a combination of building code requirements and the results of years of lessons learned related to special inspection work that have resulted in the successful completion of buildings.

B. Duties and Responsibilities of the Special Inspector

The IBC requires that a "Statement of Special Inspections" be submitted with the application of the permit. The special inspector should know and understand the scope of the statement prior to beginning special inspections (see the duties of the design professional in responsible charge).

Though not required by code, special inspectors and/or inspection agencies can document acceptance of their responsibilities and the scope of work for a project by signing an agreement that includes a detailed schedule of services, such as the Statement of Special Inspections Agreement and the Statement of Special Inspections Schedule, included in Section 1 of Appendix A.

Duties of special inspectors and/or inspection agencies include the following:

1. **General requirements.** Special inspectors shall review approved plans and specifications for special inspection requirements. Special inspectors will comply with the special inspection requirements of the enforcing jurisdiction found in the Statement of Special Inspections, including work and materials.

2. **Signify presence at job site.** Special inspectors shall notify contractor personnel of their presence and responsibilities at the job site. If required by the building official, they shall sign in on the appropriate form posted with the building permit. An example of a special inspection record is included in Section 1 of Appendix A.

3. **Observe assigned work.** Special inspectors shall inspect all work according to the Statement of Special Inspections for which they are responsible for compliance with the building-department-approved (stamped) plans and specifications, and the applicable provisions of Section 1704.

4. **Report nonconforming items (discrepancies).** Special inspectors shall bring all nonconforming items to the immediate attention of the contractor. If any such item is not resolved in a timely manner or is soon to be incorporated into the work, the design professional in responsible charge and the building official should be notified immediately and the item noted in the special inspector's written report (see Section 1704.2.4). Some jurisdictions may require this report to be a separate, individual report from the progress reports. The building official may require this

report to be posted in a conspicuous place on the job site. The special inspector should include in the report, as a minimum, the following information about each nonconforming item:

- Description and exact location.
- Reference to applicable detail of approved plans/specifications.
- Name and title of each individual notified, and method of notification.
- Resolution or corrective action taken.

5. **Provide timely progress reports.** The special inspector shall complete written inspection reports for each inspection visit and provide the reports on a timely basis as determined by the building official. The special inspector or inspection agency shall furnish these reports directly to the building official and to the design professional in responsible charge (see Section 1704.2.4). These reports should be organized on a daily format and may be submitted weekly at the option of the building official. In these reports, special inspectors should:

    - Describe inspections and tests made with applicable locations and whether the work meets the requirements of the Statement of Special Inspections.
    - Indicate nonconforming items (discrepancies) and how they were resolved.
    - List unresolved items, parties notified, and time and method of notification.
    - Itemize changes authorized by design professional in responsible charge if not included in nonconforming items.

    Examples of report forms are included in Section 2 of Appendix A.

6. **Submit final report.** Special inspectors or inspection agencies shall submit a final signed report to the building department stating that all items requiring special inspection and testing by the Statement of Special Inspection were fulfilled and reported; and, to the best of their knowledge, all items are in compliance with the approved plans and specifications (see Section 1704.2.4). Jurisdictions may also require the design professional in responsible charge to sign the report before it is submitted to the building official. Items not in conformance, unresolved items or any discrepancies in inspection coverage (i.e., missed inspections, periodic inspection when continuous was required, etc.) should be specifically itemized in this report. An example of a final report form is included in Section 2 of Appendix A.

C. Duties and Responsibilities of the Project Owner

The project owner, the design professional in responsible charge or an agent of the owner is responsible for funding special inspection services. Measures should be taken to ensure that the scope of work and duties of the special inspector as outlined in the Statement of Special Inspections are not compromised.

D. Duties and Responsibilities of the Design Professional in Responsible Charge

The design professional in responsible charge should be a consenting party by written acknowledgment of the Statement of Special Inspections agreements. An example of such an agreement is included in Section 1 of Appendix A.

The design professional in responsible charge shall prepare a "Statement of Special Inspections" to be submitted with the permit application (see Sections 107.1, 110.3.9 and 1704.2.3). The content of the statement is to contain the following information (see Section 1704.3.1).

1. Materials, systems, components and work required to have special inspection. Duties of the building official and the registered design professional responsible for each part of the work are to be stated.

Frank Williams
562-547-4564

2. Type and extent of each special inspection, and the name of individual(s) or firms performing the inspections.

3. Type and extent of each test.

4. Special inspection and testing of required seismic-resistance systems and components (see Sections 1705.11 and 1705.12).

5. Special inspection of required wind-resistance systems and components (see Section 1704.10).

6. Identification of whether each special inspection is continuous or periodic.

The following are elements of good practice and could be included in the Statement of Special Inspections.

7. Provide name and contact information of special inspector or special inspection agency. Subject to the approval of the building official, special inspectors holding current certification by ICC in the discipline in which they will be inspecting can be considered qualified. Likewise, special inspection agencies holding current accreditation with IAS (see the updated list available at www.iasonline.org) with the appropriate scope of accreditation for the disciplines to be inspected can be considered qualified. The choice of special inspectors or special inspection agencies should include the following considerations:

    - Project size and complexity – experience with similar projects.
    - Inspection staffing – sufficient qualified inspectors.
    - Site location – proximity of inspection and testing facilities.
    - Offsite inspection – capabilities for inspection at remote locations.

8. Completion of a Statement of Special Inspections agreement and schedule as shown in the examples in Appendix A is a simple method of fulfilling the requirements for preparation of a special inspection program that can be easily reviewed by the building official.

9. Respond to field discrepancies. The design professional in responsible charge shall respond to special inspector reports of uncorrected noncomplying items (discrepancies), and shall approve remedial measures.

10. Review shop drawings and submit revisions to approved plans. The design professional in responsible charge shall acknowledge and approve shop drawings that may detail structural information. The design professional shall submit to the building official and the special inspector or inspection agency written approval of any verbally approved deviations from the approved plans, and shall submit revised plans for building official approval (see Sections 107.3.4.1 and 107.4). It is becoming more common for the building official to require the signature of the design professional on all final documents.

It is good practice that the design professional in responsible charge be a signatory of the Statement of Special Inspections agreement and of any subsequent agreements affecting the work of special inspection of the project.

E. Duties and Responsibilities of the Contractor

Each contractor responsible for construction of the structural systems described in Section 1704.4 shall submit a written statement of responsibility to the building official and the owner prior to commencement of the work on the system or component. The contractor's statement of responsibility is to contain the following:

1. Acknowledgement of awareness of the special requirements contained in the Statement of Special Inspections;

2. Acknowledgement that control will be exercised to obtain compliance with the construction documents approved by the building official;

3. Procedures for exercising control within the contractor's organization, the method and frequency of reporting and the distribution of the reports; and

4. Identification and qualifications of the person(s) exercising such control and their position(s) in the organization.

In addition to the responsibilities mentioned by code, the contractor should:

5. Notify the special inspector. Adequate notice shall be provided so that the special inspector has time to become familiar with the project.

6. Provide access to approved plans. The contractor is responsible for providing the special inspector with access to approved plans (see Section 107.3.1).

7. Retain special inspection records. When required by the building official, the contractor is responsible for retaining, at the job site, all special inspection records submitted by the special inspector and providing these records for review by the building department's inspector upon request.

F. Duties and Responsibilities of the Building Official

Of all the team members involved in the construction process, the building official is the only one with the legal authority to enforce the special inspection provisions of the code (see Section 104.4). The employment of a special inspector or agency shall not relieve the building department of responsibility for inspections or structural observations as required by the code and stated in the Statement of Special Inspections. Building department inspections of items also requiring special inspection should not be signed off without the concurrence of the special inspector.

The specific duties and responsibilities of the building official relating to special inspection include the following:

1. Review submittal documents for compliance with the special inspection requirements. The building official is charged with the legal authority to review the plans, specifications pertaining to work and materials, special inspection program and other submittal documents for compliance with code requirements (see Sections 104.2 and 107). Approval for materials should be in accordance with Sections 104.9 and 104.11.

2. Approve fabricator(s) used for building components installed on-site.

3. Approve special inspection program. The building official is responsible for approving the special inspection program submitted by the design professional in responsible charge (see Section 107.3.4) and may require a preconstruction conference to review the program with all applicable members of the construction team.

4. Approve special inspectors/inspection agencies. The building official is responsible for determining the competence of special inspectors for the types of work they will be inspecting (see Section 1704) and requiring special inspectors/inspection agencies to notify the building department of each permit they are associated with. Suggested qualifications for special inspectors are contained in Part III of this guide.

5. Monitor special inspection activities. The building official should monitor the special inspection activities at the job site to ensure that qualified special inspectors are performing their duties when work requiring special inspection is in progress (see Section 1704.2.4).

6. Issuance of stop work orders. The building official is recognized as having the authority to stop work at the job site (see Section 115); however, the special inspector does not have this authority. Stop work orders most likely will be issued in response to a discrepancy report that has been written.

7. Approval to proceed. There are certain points of completion where work shall not proceed until approval by the building official has been given.

8. Review inspection reports. The building official receives and reviews special inspection progress reports and final reports for compliance with the approved plans, specifications and workmanship provisions of the code (see Section 1704.2.4).

9. The building official should not perform the final inspection and issue certificate of occupancy for a project (see Section 111) until the final special inspection report has been reviewed and approved.

(This page is intentionally left blank.)

III. RECOMMENDED SPECIAL INSPECTOR QUALIFICATIONS

A. Code Knowledge and Plan Reading Competency Standards

The building official has the responsibility to ascertain that special inspectors and/or agencies are qualified for the type(s) of inspection required. In some geographic areas, cooperative jurisdictional programs for the recognition of competent inspectors/agencies are available to assist in satisfying this requirement. Qualification standards may include:

1. Individuals maintaining current certification by ICC as a special inspector for the discipline(s) for which he or she is requesting approval.

2. Agencies maintaining current accreditation as a special inspection agency by IAS with a scope of accreditation covering the discipline(s) for which the agency is requesting approval. See Appendix C for the IAS-required minimum qualifications for special inspection.

3. Written or oral examination to verify the applicant's knowledge of jurisdictional procedures and code requirements.

B. Experience Standards

The specialized knowledge and skills needed to perform special inspection should be obtained by experience in some aspect of the discipline being inspected. Industry standards typically require one to five years of verifiable practical experience, depending upon the discipline and the level of inspection, a portion of which may be satisfied by applicable technical education. There are also provisions for trainees who may work under the direct supervision of qualified special inspectors. An example of detailed guidelines for special inspector experience standards is contained in Appendix C.

The building official may require that special inspector applicants provide evidence of experience through the following and/or other means:

1. Written references verifying related work experience; and/or

2. Personal interview to evaluate the applicant's work experience and suitability to be a special inspector.

3. Evidence of compliance with a recognized industry experience guideline (see examples in Appendix C).

C. Performance Standards

The building official may use the job task lists for special inspector certification exams as guidelines for evaluating the performance of special inspectors in each of the applicable disciplines. These lists are included in Appendix B, and contain comprehensive tasks that may be performed by special inspectors on large and complex projects. Actual tasks required on specific projects will typically be fewer.

(This page is intentionally left blank.)

IV. RECOMMENDED TESTING AND INSPECTION AGENCIES ACCREDITATION

A. Special Inspection Agencies Accreditation Program

In response to a request from several ICC governmental members, IAS has developed an accreditation program for agencies providing the special inspections that are required for specific construction projects under Chapter 17 of the 2012 IBC. The IAS Special Inspection Agency Accreditation Program is based primarily on the requirements of the IBC and the applicable portions of ISO/IEC 17020, *General Criteria for the Operation of Various Bodies Performing Inspection*. The program requires special inspection agencies to operate under a quality management system that is documented in a manual, and also requires the agency to be assessed in the field to determine if it is competent to perform specific inspections or types of inspections. IAS accreditation is based on the assessment of a special inspection agency's inspection procedures, the competence of its inspection staff and its reporting procedures.

B. Testing Laboratories Accreditation Program

Industry depends on reliable assessment of testing and calibration laboratories, inspection agencies and fabricator inspection programs. Accurate test, calibration and inspection results are vital for the protection of public health and safety, and to facilitate trade. IAS operates nationally and internationally recognized accreditation programs, some of which are specifically tailored to the needs of the building official. The IAS laboratory accreditation program is modeled on ISO/IEC 17025 and is recognized globally by the International Laboratory Accreditation Cooperation (ILAC), and by the Asia Pacific Laboratory Accreditation Cooperation (APLAC). Laboratory accreditation, traceability of measurements and uniform testing requirements minimize technical barriers to trade and reduce the financial burden on manufacturers. Evaluation of a product, material or piece of equipment should be based on reports from accredited testing laboratories that are supported by accredited calibration laboratories to ensure traceable and repeatable measurements.

C. Public, Affected Parties and Regulator Access

The IAS accreditation programs are monitored by an accreditation committee that is composed of code regulators who approve all accreditation criteria.Through its accreditation committee, IAS follows a public hearing process that allows for input from regulators, laboratories, inspection agencies and other interested parties. Members of the accreditation committee are appointed by the IAS Board of Directors. The accreditation committee provides a transparent process that provides access to all interested parties. On technical matters IAS maintains formally appointed Technical Advisory Councils (TACs) consisting of experts drawn from accredited calibration and testing laboratories, and special inspection agencies. Each TAC provides technical guidance to IAS on matters relating to calibration and testing or special inspection as the need arises.

Requiring accreditation is an alternative that can significantly reduce unnecessary oversight burden on a building department. Through this comprehensive program, a building department has the assurance that special inspection agencies have fully demonstrated that their experience and knowledge is sufficient, and that they are competent to determine if construction meets the material and workmanship requirements of the jurisdiction's contract documents and codes.

The steps recommended to implement a program requiring IAS accreditation for special inspection agencies are:

1. Drafting a jurisdictional ordinance requiring special inspection agencies to become accredited to work in the jurisdiction.
2. Establishing a timeline for the agencies to complete the accreditation process.
3. Educating special inspection agencies on the program and the accreditation process.
4. Educating design professionals and contractors on the effective date of compliance.

Special inspection administration is one of the areas evaluated by IAS for accreditation of building departments, fire prevention departments and code enforcement agencies.

APPENDIX A – SECTION 1

Examples of Special Inspection Forms, Schedules and Agreements
Municipal Agency Forms

- Statement of Special Inspections Agreement
- Statement of Special Inspections Schedule
- Special Inspection Record

(THESE FORMS MAY BE PHOTOCOPIED)

(This page is intentionally left blank.)

STATEMENT OF SPECIAL INSPECTIONS AGREEMENT

To permit applicants of projects requiring special inspection and/or testing in accordance with Section 1704 of the *International Building Code* (IBC):

Project Address: ______________________________ **Permit No.:** ________________________

BEFORE A PERMIT CAN BE ISSUED, two (2) copies of this agreement, including the Statement of Special Inspections and the Staement of Special Inspections Schedule with the required acknowledgments, shall be completed by the owner or the registered design professional in responsible charge acting as the owner's agent. A preconstruction conference with the parties involved may be required to review the special inspection requirements and procedures.

APPROVAL OF SPECIAL INSPECTORS: Special inspectors may have no financial interest in projects for which they provide special inspection. Special inspectors shall be approved by the building department prior to performing any duties. Special inspectors shall submit their qualifications and are subject to personal interviews for prequalification. Special inspectors shall display approved identification, as stipulated by the building official, when performing the function of special inspector.

Special inspection and testing shall meet the minimum requirements of Sections 1704 and 1705 of the IBC. The following conditions also are applicable:

A. Duties and Responsibilities of the Special Inspector

1. **General requirements.** Special inspectors shall review approved plans and specifications for special inspection requirements. Special inspectors will comply with the special inspection requirements of the enforcing jurisdiction found in the Statement of Special Inspections, including work and materials.

2. **Signify presence at job site.** Special inspectors shall notify contractor personnel of their presence and responsibilities at the job site. If required by the building official, they shall sign in on the appropriate form posted with the building permit.

3. **Observe assigned work.** Special inspectors shall inspect all work according to the Statement of Special Inspections for which they are responsible for compliance with the building-department-approved (stamped) plans and specifications, and the applicable provisions of Section 1704 of the IBC.

4. **Report nonconforming items (discrepancies).** Special inspectors shall bring all nonconforming items to the immediate attention of the contractor. If any such item is not resolved in a timely manner or is about to be incorporated into the work, the design professional in responsible charge and the building official should be notified immediately and the item noted in the special inspector's written report (see Section 1704.2.4). The building official may require this report to be posted in a conspicuous place on the job site. The special inspector should include in the report, as a minimum, the following information about each nonconforming item:

    - Description and exact location.
    - Reference to applicable detail of approved plans/specifications.
    - Name and title of each individual notified and method of notification.
    - Resolution or corrective action taken.

5. **Provide timely progress reports.** The special inspector shall complete written inspection reports for each inspection visit and provide the reports on a timely basis as determined by the building official. The special inspector or inspection agency shall furnish these reports directly to the building official and to the design professional in responsible charge (see Section 1704.2.4). These reports should be organized on a daily format and may be submitted weekly at the option of the building official. In these reports, special inspectors should:

   - Describe inspections and tests made with applicable locations.
   - Indicate nonconforming items (discrepancies) and how they were resolved.
   - List unresolved items, parties notified, and time and method of notification.
   - Itemize changes authorized by the design professional in responsible charge if not included in nonconforming items.

6. **Submit final report.** Special inspectors or inspection agencies shall submit a final signed report to the building department stating that all items requiring special inspection and testing by the Statement of Special Inspections were fulfilled and reported, and, to the best of their knowledge, conform to the approved plans and specifications (see Section 1704.2.4). Some jurisdictions also require the design professional in responsible charge to sign the report before it is submitted to the building official. Items not in conformance, unresolved items or any discrepancies in inspection coverage (i.e., missed inspections, periodic inspection when continuous was required, etc.) should be specifically itemized in this report.

B. Owner Responsibilities

The project owner, the design professsional in responsible charge or an agent of the owner is responsible for funding special inspection services. Measures should be taken to ensure that the scope of work and duties of the special inspector as outlined in the Statement of Special Inspections are not compromised.

C. Registered Design Professional in Responsible Charge

The design professional in responsible charge should be a consenting party by written acknowledgment of special inspection and testing agreements. The design professional in responsible charge has many duties and responsibilities related to special inspection, including the following:

The Statement of Special Inspections must be submitted with the permit application. The content of the statement contains the following information.

1. Materials, systems, components and work required to have special inspection. Duties of the building official and the registered design professional responsible for each part of the work are to be stated.

2. Type and extent of each special inspection, and the name of individuals or firms performing the inspections.

3. Type and extent of each test.

4. Special inspection of required seismic-resistant systems and components.

5. Special inspection of required wind-resistant systems and components.

6. Provide name and contact information of special inspector or special inspection agency. Subject to the approval of the building official, special inspectors holding current certification by ICC in the discipline in which they will be inspecting can be considered qualified. IAS-accredited spe-

cial inspection agencies providing inspections within the scope of accreditation for the disciplines to be inspected can be considered qualified. The choice of special inspectors or special inspection agencies should include the following considerations:

- Project size and complexity – experience with similar projects.
- Inspection staffing – sufficient qualified inspectors.
- Site location – proximity of inspection and testing facilities.
- Off-site inspection – capabilities for inspection at remote locations.

7. Completion of a Statement of Special Inspections agreement and schedule as shown in this appendix is a simple method of fulfilling the requirement for preparation of a special inspection program that can be easily reviewed by the building official.

8. Respond to field discrepancies. The design professional in responsible charge shall respond to special inspector reports of uncorrected noncomplying items (discrepancies) and shall approve remedial measures.

9. Review shop drawings and submit revisions to approved plans. The design professional in responsible charge shall acknowledge and approve shop drawings that may detail structural information. The design professional shall submit to the building official and the special inspector/inspection agency written approval of any verbally approved deviations from the approved plans, and shall submit revised plans for building official approval (see Sections 107.3.4 and 107.4).

D. Contractor Responsibilities

Submit a written statement of responsibility to the building official and the owner prior to commencement of the work on the system or component. The contractor's statement of responsibility is to contain the following:

1. Acknowledgement of awareness of the special requirements contained in the Statement of Special Inspections;

2. Acknowledgement that control will be exercised to obtain compliance with the construction documents approved by the building official;

3. Procedures for exercising control within the contractor's organization, the method and frequency of reporting, and the distribution of the reports; and

4. Identification and qualifications of the person(s) exercising such control and their position(s) in the organization.

Suggested job-site protocol for special inspection:

1. Notify the special inspector. Adequate notice shall be provided so that the special inspector has time to become familiar with the project.

2. Provide access to approved plans. The contractor is responsible for providing the special inspector with access to approved plans.

3. Retain special inspection records. When required by the building official, the contractor is responsible for retaining, at the job site, all special inspection records submitted by the inspector and providing these records for review by the building department's inspector upon request.

E. Building Department Responsibilities

1. Review submittal documents for compliance with special inspection requirements as outlined in the Statement of Special Inspections. The building official is charged with the legal authority to review the plans, specifications, special inspection program and other submittal documents for compliance with code requirements.

2. Approve fabricator(s) used for building components installed on-site.

3. Approve special inspection programs. The building official is responsible for approving the special inspection program submitted by the design professional in responsible charge (see Section 107.1) and may require a preconstruction conference to review the program with all applicable members of the construction team.

4. Monitor special inspection activities. The building official should monitor the special inspection activities at the job site to ensure that qualified special inspectors are performing their duties when work requiring special inspection is in progress.

5. Issuance of stop work orders. The building official is recognized as having the authority to stop work at the job site.

6. Approval to proceed. There are certain points of completion where work shall not proceed until approval by the building official has been given.

7. Review inspection reports. The building official receives and reviews special inspection progress reports and final reports for compliance with the approved plans, specifications and workmanship provisions of the code.

8. Perform final inspection. The building official should not perform the final inspection and approval for a project until the final special inspection report has been reviewed and approved.

ACKNOWLEDGMENTS

I have read and agree to comply with the terms and conditions of this agreement.

Owner:

______________________________ By: _________________________ Date: ______________

Registered Design Professional in Responsible Charge:

______________________________ By: _________________________ Date: ______________

Contractor:

______________________________ By: _________________________ Date: ______________

Special Inspector/Special Inspection Agency*:

______________________________ By: _________________________ Date: ______________

Others as required by Building Official:

______________________________ By: _________________________ Date: ______________

ACCEPTED FOR THE BUILDING DEPARTMENT

By: ____ Date: ______________

* This signature may be that of the Responsible Professional Engineer within the special inspection agency.

(This page is intentionally left blank.)

STATEMENT OF SPECIAL INSPECTIONS SCHEDULE

(This schedule should match the Statement of Special Inspections)

Where the option between continuous and periodic inspection is possible, circle the option required.

Steel Construction – Verification/Inspection See Table 1705.2.2 – Required Verification and Inspection of Steel Construction Other Than Structural Steel	**Continuous**	**Periodic**
1. Material verification of cold-formed steel deck:		
a. For structural steel, identification markings to conform to AISC 360.	—	X
b. For other steel, identification markings to conform to ASTM standards specified in the approved construction documents.	—	X
c. Manufacturer's certified test reports.	—	X
2. Inspection of welding:		
a. Cold-formed steel deck:		
1) Floor and roof deck welds.	—	X
b. Reinforcing steel:		
1) Verification of weldability of reinforcing steel other than ASTM A 706.	—	X
2) Reinforcing, steel-resisting flexural and axial forces in intermediate and special moment frames, and boundary elements of special structural walls of concrete and shear reinforcement.	X	—
3) Shear reinforcement.	X	—
4) Other reinforcing steel.	—	X

Concrete Construction – Verification/Inspection See Table 1705.3 – Required Verification and Inspection of Concrete Construction	**Continuous**	**Periodic**
1. Inspection of reinforcing steel welding in accordance with Table 1705.2.2, Item 2b.	—	X
2. Inspection of anchors cast in concrete where allowable loads have been increased or where strength design is used.	—	X
3. Inspection of anchors post-installed in hardened concrete members.	—	X
4. Verifying use of required design mix.	—	X
5. At the time fresh concrete is sampled to fabricate specimens for strength tests, perform slump and air content tests, and determine the temperature of the concrete.	X	—
6. Inspection of concrete and shotcrete placement for proper application techniques.	X	—
7. Inspection for maintenance of specified curing temperature and techniques.	—	X
8. Inspection of prestressed concrete:		
a. Application of prestressing forces.	X	—
b. Grouting of bonded prestressing tendons in the seismic-force-resisting system.	X	—

(continued)

Concrete Construction – Verification/Inspection—continued See Table 1705.3 – Required Verification and Inspection of Concrete Construction	**Continuous**	**Periodic**
9. Erection of precast concrete members.	—	X
10. Verification of in-situ concrete strength, prior to the stressing of tendons in post-tensioned concrete, and prior to the removal of shores and forms from beams and structural slabs.	—	X
11. Inspect formwork for shape, location and dimensions of the concrete member being formed.	—	X

Soils – Verification/Inspection See Table 1705.6 – Required Verification and Inspection of Soils	**Continuous**	**Periodic**
1. Verify materials below shallow foundations are adequate to achieve the design-bearing capacity.	—	X
2. Verify excavations are extended to proper depth and have reached proper material.	—	X
3. Perform classification and testing of compacted fill materials.	—	X
4. Verify use of proper materials, densities and lift thicknesses during placement and compaction of compacted fill.	X	—
5. Prior to placement of compacted fill, observe subgrade and verify that site has been prepared properly.	—	X

Deep Driven Foundation Elements – Verification/Inspection See Table 1705.7 – Required Verification and Inspection of Deep Driven Foundation Elements	**Continuous**	**Periodic**
1. Verify element materials, sizes and lengths comply with the requirements.	X	—
2. Determine capacities of test elements and conduct additional load tests, as required.	X	—
3. Observe driving operations and maintain complete and accurate records for each element.	X	—
4. Verify placement locations and plumbness; confirm type and size of hammer; record number of blows per foot (mm) of penetration; determine required penetrations to achieve design capacity; record tip and butt elevations; and document any damage to foundation element.	X	—
5. For steel elements, perform additional inspections in accordance with Section 1705.2.	—	—
6. For concrete elements and concrete-filled elements, perform additional inspections in accordance with Section 1705.3.	—	—
7. For specialty elements, perform additional inspections, as determined by the registered design professional in responsible charge.	—	—

Cast-in-place Deep Foundation Elements – Verification/Inspection See Table 1705.8 – Required Verification and Inspection of Cast-in-Place Deep Foundation Elements	**Continuous**	**Periodic**
1. Observe drilling operations, and maintain complete and accurate records for each element.	X	—
2. Verify placement locations and plumbness; confirm element diameters; bell diameters (if applicable); lengths; embedment into bedrock (if applicable); and adequate end-bearing strata capacity. Record concrete or grout volumes.	X	—
3. For concrete elements, perform additional inspections in accordance with Section 1705.3.	—	—

Other Materials – Verification/Inspection		

Other Work – Verification/Inspection		

Note: As schedules are developed, space for the special inspector, date of inspection and the design profession in responsible charge should be provided, as well as others who may need to recognize that the inspections have been completed.

(This page is intentionally left blank.)

SPECIAL INSPECTION RECORD

Inspection Agency: ______________________

City of ________________________________ **Project Address:** ______________________

Building Permit No.: ____________________

Note: Each special inspector is to record their presence on the job site for each day's inspection. Post this card adjacent to the building permit inspection report card. Weekly reports are to be submitted by each special inspector/inspection agency to the building department.

When attached to the job inspection record card, this card becomes a part of the inspection record.

INSPECTION TYPE	SPECIAL INSPECTOR	ID NO.	DATE	NOTES	TIME	
					ARRIVAL	DEPARTURE

(This page is intentionally left blank.)

APPENDIX A – SECTION 2

Examples of Special Inspection Forms, Schedules and Agreements
Special Inspection Forms

- Special Inspection Daily Report
- Special Inspection Weekly Report
- Special Inspection Discrepancy Notice
- Special Inspection Final Report

(THESE FORMS MAY BE PHOTOCOPIED)

(This page is intentionally left blank.)

SPECIAL INSPECTION DAILY REPORT

City/County of ______________________________ Permit No.: ___________ Date: _______________

Project Name/Address: ____

Inspection Type(s)/Coverage: ____

____

☐ Continuous ☐ Periodic; frequency: ____________________________

Inspections made, including locations: ____

____

____

Tests performed: ____

____

____

Items requiring (1) correction, (2) correction of previously listed items and (3) previously listed uncorrected items: ____

____

____

Changes to approved plans authorized by the design professional in responsible charge: ______

____

____

Comments: ____

____

To the best of my knowledge, the work inspected was in accordance with the building-department-approved plans, specifications and applicable workmanship provisions of the IBC, except as noted above.

Signed: _________________________________ Inspection Agency*: ________________________

Print full name: ___________________________ ID/Certificate Number: ____________________

* Building official may require the signature/stamp of agency engineer responsible for special inspection.

(This page is intentionally left blank.)

SPECIAL INSPECTION WEEKLY REPORT

City/County of ______________________ Permit No.: __________ Date: ______________

Project Name/Address: ____

Inspection Type(s)/Coverage: ____

____

☐ Continuous ☐ Periodic; frequency: ______________________

Total inspection time each day:

Dates							
Hours							
Inspector							

Inspections made, including locations: ____

____

Tests performed: ____

____

Items requiring (1) correction, (2) correction of previously listed items and (3) previously listed uncorrected items: ____

____

Changes to approved plans authorized by the design professional in responsible charge: ______

____

Comments: ____

____

To the best of my knowledge, the work inspected was in accordance with the building-department-approved plans, specifications and applicable workmanship provisions of the IBC, except as noted above.

Signed: ______________________ Inspection Agency*: ______________________

Print full name: ______________________ ID/Certificate Number: ______________________

cc: Building Department
Engineer/Architect

* Building official may require the signature/stamp of agency engineer responsible for special inspection.

(This page is intentionally left blank.)

SPECIAL INSPECTION DISCREPANCY NOTICE

City/County of ______________________ **Permit No.:** __________ **Date:** ______________

Project Name/Address: ____

Inspection Type(s)/Coverage: ____

____

☐ **Continuous** ☐ **Periodic; frequency:** ______________________

Notice delivered to: ☐ **Contractor** ☐ **Engineer/Architect** ☐ **Building Department**

The following discrepancies require correction and inspection approval prior to proceeding with this phase of the work:

____

____

____

____

____

____

____

____

____

____

Signed: ______________________ **Inspection Agency*:** ______________________

Print full name: ______________________ **ID/Certificate Number:** ______________________

*** Building official may require the signature/stamp of agency engineer responsible for special inspection.**

DO NOT REMOVE THIS NOTICE

Post with building permit inspection record card

(This page is intentionally left blank.)

SPECIAL INSPECTION FINAL REPORT

City/County of ______________________ Permit No.: __________ Date: ____________

Attention: ____________________________

Project Name/Address: ____

____

In accordance with Section 1704 of the *International Building Code*, special inspection has been provided for the following items:

____

____

____

____

____

____

____

____

____

____

Based upon inspections performed and our (my) substantiating reports, it is our (my) professional judgment that, to the best of our (my) knowledge, the inspected work was performed in accordance with the approved plans, specifications and applicable workmanship provisions of the *International Building Code*.

Signed: __________________________ Inspection Agency*: ____________________

Print full name: ____________________ ID/Certificate Number: ____________________

cc: Building Department
Engineer/Architect

* Building official may require the signature/stamp of agency engineer responsible for special inspection.

(This page is intentionally left blank.)

APPENDIX B

Job Task Lists for Special Inspectors

- Reinforced Concrete
- Prestressed Concrete
- Structural Masonry
- Structural Steel and Bolting
- Structural Welding
- Spray-applied and Intumescent Fireproofing

(This page is intentionally left blank.)

JOB TASK LISTS FOR SPECIAL INSPECTORS

These task lists are the basis of the ICC certification exams. Information for the exams is found in the 2011 National Certification Examination Program Bulletin found at: http://www.iccsafe.org/Accreditation/Bulletins/National_Certification_EIB.pdf.

The job task lists are developed by the exam development committee specifically assigned to each exam. The tasks describe the work performed by special inspectors as it is practiced in the field. Other tasks related to this work may be helpful or extend the professional ability of the inspector; however, these tasks are the minimum required to protect the public health and safety. Minimum requirements for certification exams are mandated by the Equal Employment Opportunity Commission (EEOC), and accepted professional certification standards as developed by the American Psychological Association (APA), Council on Licensure, Enforcement and Regulation (CLEAR), and National Organization of Competency Assurance (NOCA).

The building official and others may desire to review these tasks in reviewing the proposed special inspection activities for a particular project and evaluating the qualifications of special inspectors. Also, the minimum qualifications listed in IAS AC291, *Accreditation Criteria for Special Inspection Agencies*, outlined in Appendix C, provides examples of qualifications.

A. Reinforced Concrete Special Inspector

1. General requirements

Review the approved plans and specifications for special inspection requirements. Comply with special inspection requirements of the enforcing jurisdiction. Notify the contractor of deviations from approved plans and specifications. If the deviations are uncorrected, notify the design professional in responsible charge and the building official. Submit progress reports to the design professional and the building official describing the tests that were performed and the compliance of the work. Submit a final summary report stating whether the work requiring special inspection was in compliance with the approved plans and applicable provisions of the IBC.

2. Concrete quality

Verify that the individual batch tickets indicate delivery of the approved mix, as specified. Verify the time limits of mixing, total water added, and proper consistency and workability for placement. Determine the required type, quantity and frequency of tests to be performed on fresh and hardened concrete. Observe the sampling of concrete, field testing of fresh concrete and making of test specimens. Provide or arrange for the proper specimen identification, site storage and protection, and transportation to the testing laboratory. Provide or arrange for the communication of field-testing results to the design professional in responsible charge and the building official.

3. Reinforcement

Verify that the reinforcing steels are of the type, grade and size specified, and are in compliance with acceptable quality standards. Ensure that the reinforcing steel is free of oil, dirt and rust, and that the steel is properly coated and/or sheathed, as specified. Verify that the reinforcing steels are located within acceptable tolerances, and are adequately supported and secured to prevent displacement during concrete placement. Verify that minimum concrete cover is provided. Verify that the placement of reinforcing steel (or ducts) complies with required spacing, profile and quantity requirements, as indicated by both the approved plans and installation drawings. Verify that hooks, bends, ties, stirrups and supplemental reinforcement are fabricated and placed, as specified. Verify that the required lap lengths, stagger and offsets are provided. Verify the proper installation of approved mechanical connections in accordance with the manufacturer's instructions and/or evaluation reports. Ensure that all welds of reinforcing steel and other weldments are as specified, and have been inspected and approved by an approved welding inspector.

4. Formwork, joints and embeds

Verify that the formwork will provide concrete elements of the specified size and shape. Verify that the location and preparation of construction joints are in accordance with the approved plans, specifications and building code requirements. Verify that the type, quantity, size, spacing and location of embedded items are as specified.

5. Concrete placement, protection and curing

Verify the acceptable condition of the place of deposit before the concrete is placed. Verify that the methods of conveying and depositing concrete avoid contamination and segregation of the mix. Verify that the concrete is being properly consolidated during placement. Verify that the concrete is protected from temperature extremes, and determine that the proper curing is initiated.

B. Prestressed Concrete Special Inspector

Note: Certification as a Reinforced Concrete Special Inspector is required before certification as a Prestressed Concrete Special Inspector. All items listed under "Reinforced Concrete Special Inspector" are considered prerequisite to the knowledge for special inspection of prestressed concrete.

1. General requirements

Comply with the special inspection requirements of the enforcing jurisdiction. Review approved plans and specifications for project details that pertain to special inspection requirements. Notify the contractor of deviations from approved plans and specifications. If the deviations are uncorrected, notify the design professional in responsible charge and the building official. Submit progress reports to the design professional and the building official describing the tests that were performed and the compliance of the work. Submit a final summary report stating whether the work requiring special inspection was in compliance with the approved construction documents and applicable provisions of the IBC.

2. Concrete quality

Verify that the individual batch tickets indicate delivery of the approved mix, as specified. Verify the time limits of mixing, total water added, and proper consistency and workability for placement. Determine the required type, quantity and frequency of tests to be performed on fresh and hardened concrete. Observe the sampling of concrete, field testing of fresh concrete and making of test specimens. Provide or arrange for the proper specimen identification, site storage and protection, and transportation to the testing laboratory. Provide or arrange for the communication of field-testing results to the design professional in responsible charge and the building official.

3. Reinforcement

Verify that reinforcing steel and tendons are of the type, grade and size specified, and are in compliance with acceptable quality standards. Verify that the reinforcing steel and tendon system are fabricated in compliance with acceptable quality standards. Verify that the tendons at the time of concrete placement are free of oil, dirt and excessive rust, and are properly coated and/or sheathed, as specified. Verify that the reinforcing steel and tendons are located within specified tolerances, and are adequately supported and secured to prevent displacement during concrete placement. Verify that minimum concrete cover is provided. Verify that the placement of reinforcing steel and tendons (or ducts) comply with spacing, profile and quantity requirements, as indicated by the installation drawings and approved plans. Verify that hooks, bends, ties, stirrups and supplemental reinforcement are fabricated and placed, as specified. Verify that the required lap lengths, stagger and offsets are provided. Verify the proper installation of approved mechanical connections in accordance with the manufacturer's instructions and/or evaluation reports. Verify that welds have been inspected and approved, as specified. Verify that prestressed rock and soil anchors are fabricated and installed, as specified.

4. Prestressing and grouting

Verify that the required concrete strength has been attained prior to transferring prestressing forces. Verify the proper equipment calibration. Verify that the proper stressing (or tensioning) sequences are used, proper jacking forces are applied, and acceptable elongations are attained and recorded. Verify that tendons and anchorages are properly sealed or otherwise protected, as specified. Verify that ducts, including inlets and outlets, are of the required size, are mortar-tight and are located correctly. Verify that the proper grout materials, strength and grouting pressures are used, as specified.

5. Formwork, joints and embedments

Verify that the formwork will provide concrete elements of the specified size and shape. Verify that the location and preparation of construction joints are in accordance with the approved plans, specifications, and applicable codes and standards. Verify that the type, quantity, size, spacing, condition and location of embedded items are as specified.

6. Concrete placement, protection and curing

Verify the acceptable condition of the place of deposit before the concrete is placed. Verify that the methods of conveying and depositing concrete avoid contamination; segregation of the mix; and displacement of reinforcement, embedments and forms. Verify that the concrete is being properly consolidated during placement. Verify that the concrete is protected from ambient temperature extremes during placement and curing. Verify that the concrete is being cured as specified by approved plans, specifications and applicable codes.

C. Structural Masonry Special Inspector

1. General requirements

Review the approved plans, specifications and submittals for special inspection requirements. Comply the with special inspection requirements of the enforcing jurisdiction in accordance with the applicable code, approved plans and specifications. Notify the contractor of deviations from the approved plans and specifications. If the deviations are uncorrected, notify the design professional in responsible charge and the building official of deviations. Submit progress reports to the design professional and the building official describing the tests that were performed and the compliance of the work. Submit a final summary report stating whether the work requiring special inspection was in compliance with the approved plans and applicable provisions of the IBC.

2. Materials

Verify that brick, block, cement, lime, aggregates, reinforcement, connectors, water, admixtures and other materials are of the type specified and approved. Verify that materials are properly stored. Verify that the mix proportions, material handling and mixing are in accordance with code requirements. Verify that the grout is batched in accordance with the approved mix. Determine the required material strengths, type and frequency of tests to be performed. Observe sampling, field testing and fabrication of test specimens. Verify that the masonry strength meets approved specifications. Verify proper sample identification, site storage, protection and transportation to the testing laboratory.

3. Masonry placement

Verify that the condition of the substrate is acceptable for placement, that mortar is properly placed and that the masonry units are placed in accordance with the approved plans. Verify that the type, quantity, size, spacing and location of embedded items are as specified. Verify that the location and preparation of movement joints are in accordance with the approved plans, specifications and building code requirements. Verify that the masonry is protected from temperature extremes and adverse weather conditions.

4. Reinforcement and connector placement

Verify that the reinforcing steel and connectors comply with the required size, spacing, profile, condition and quantity requirements, as indicated by both the approved plans and installation drawings. Verify that the reinforcing steel and connectors are placed in the proper location within acceptable tolerances. Verify minimum coverage and clearance to masonry surfaces. Verify that hooks, bends, ties, stirrups and supplemental reinforcement are fabricated and placed, as specified. Verify that the required lap lengths, stagger and offsets are provided. Verify the installation of approved mechanical connections in accordance with the manufacturer's instructions and/or evaluation reports.

5. Grout placement

Verify that grout spaces are free of obstructions and that cleanouts are provided, as required. Verify that the methods of conveying and placing grout avoid contamination and segregation, and comply with time limits and grout-lift requirements. Verify that the grout is being properly consolidated and reconsolidated during placement.

D. Structural Steel and Bolting Special Inspector

1. General requirements

Review the approved plans and specifications for special inspection requirements. Comply with the special inspection requirements of the enforcing jurisdiction. Notify the contractor of deviations from the approved plans and specifications. If the deviations are uncorrected, notify the design professional in responsible charge and the building official of deviations. Submit progress reports to the design professional and the building official describing the tests that were performed and the compliance of the work. Submit a final summary report stating whether the work requiring special inspection was in compliance with the approved plans and applicable provisions of the IBC.

2. Material sampling, testing and verification

Verify that the steel shapes and bolts are of the type, size, grade and condition specified on the approved plans and specifications. Verify the required type, quantity, location and frequency of the tests to be performed, and witness the preparation of properly identified test samples on all materials. Provide or arrange for the documentation and transportation of samples to the laboratory. Verify that the required testing is performed on materials as required by applicable standards and specifications.

3. High-strength bolting

Verify the correct type, size and location of the bolts and bolt holes, nuts and washers for the type of connection specified on approved plans and specifications. Verify the protected storage of bolts, nuts and washers as required by applicable standards and specifications. Verify that the faying surfaces at connections utilizing high-strength bolts are in compliance with applicable standards. Observe or conduct bolt-tension verification tests on required high-strength bolt assemblies. Identify and verify the joint type and installation of bolt assemblies in accordance with approved plans and specifications. Verify the use of the approved method and sequence of bolt tightening.

4. Steel framing observation

Verify that the structural steel frame orientation, details and frame member sizes are in accordance with approved plans and specifications. Verify that the column base plates are of the designed configuration, have the correct size hole and have proper clearance for grouting. Verify the grout placement and sampling. Verify that the base plates are securely seated and fastened in accordance with applicable plans and specifications.

E. Structural Welding Special Inspector

Note: The Structural Steel and Bolting Special Inspector Certification is a prerequisite for receiving the certification for Structural Welding Special Inspector.

1. General requirements

Review the approved plans and specifications for the special inspection requirements. Comply with the special inspection requirements of the enforcing jurisdiction. Notify the contractor of deviations from the approved plans and specifications. If the deviations are uncorrected, notify the design professional in responsible charge and the building official. Submit progress reports to the design professional and the building official describing the tests that were performed and the compliance of the work. Submit a final summary report stating whether the work requiring special inspection was in compliance with the approved plans and applicable provisions of the IBC.

2. Material sampling, testing and verification

Verify that the steel shapes, base metals, filler metals and gases are of the type, size, grade and condition specified on the approved plans, specifications and welding procedures specifications. Verify the required type, quantity, location and frequency of tests to be performed, and witness the preparation of properly identified test samples on all materials. Provide or arrange for the documentation and transportation of samples to the laboratory. Verify that required destructive testing is performed on materials, as required by applicable standards and specifications. Verify that required nondestructive examinations are performed, as required by applicable standards and specifications.

3. Structural, reinforcing and sheet steel welding

Verify that the welding equipment and process has the capability to produce the specified welds. Ensure that the welding equipment is calibrated and appropriate for use with the welding process. Verify and/or witness the qualification of welders, welding operators and tackers for compliance with American Welding Society (AWS) standards and specifications. Verify that welders are qualified to perform the specified work. Verify that the proposed welding procedure for structural steel, reinforcing steel and sheet metal is a standard prequalified procedure, or has been properly qualified and approved. Verify that the welding processes, sequences and procedures are followed in accordance with approved welding procedures specifications. Review the approved plans and specifications for weld types and locations. Verify that the filler materials are stored and handled in accordance with the manufacturer's and project specifications. Verify that the base metal to be welded is properly prepared and oriented. Verify that weldments have the proper joint geometry, and have backing and start/runoff tabs where required. Inspect to ensure that weld and structural steel repairs are performed in accordance with approved procedures. Verify that the fabricated elements are within permissible tolerances. Verify that welds have the specified length and effective throat. Verify that the weld profile meets applicable shape, size and quality requirements.

F. Spray-applied and Intumescent Fireproofing

1. General requirements

Review the approved plans and specifications for the special inspection requirements. Comply with the special inspection requirements of the enforcing jurisdiction. Notify the contractor of deviations from approved plans and specifications. If the deviations are uncorrected, notify the design professional in responsible charge and the building official. Submit progress reports to the design professional and the building official describing the tests that were performed and the compliance of the work. Submit a final summary report stating whether the work requiring special inspection was in compliance with the approved plans and applicable provisions of the IBC.

2. Materials, preparation, application and testing

Verify that the proposed materials of the type specified are properly stored and have been approved by the registered design professional in responsible charge and the building official. Verify that the substrate has been properly prepared; and is free of oil, dirt, scale, loose paint or primer, and other materials that may prevent adequate adhesion. Identify the members to be fireproofed, and the minimum required coverage and thickness. Verify the condition of the finished application. Determine the required type and frequency of the tests to be performed. Observe the sampling, field testing and fabrication of the test specimens. Verify that the materials are of the type specified, properly stored and approved; verify that the substrate has been properly prepared and is free of conditions that may prevent adhesion; identify the members to be fireproofed, the minimum required coverage and thickness of the fireproofing, and the condition of the finished application; and determine the required tests and observation of the sampling, field testing and fabrication of test specimens.

3. Soils

Special inspections for existing-site soil conditions, fill placement and load-bearing requirements shall be as required in Section 1705.6 and Table 1705.6.

Note the exception regarding Section 1803.

(This page is intentionally left blank.)

APPENDIX C

Examples of Special Inspector Qualifications
In Accordance with IAS AC291

(This page is intentionally left blank.)

SPECIAL INSPECTOR QUALIFICATIONS

The minimum qualifications listed are from IAS AC291, *Accreditation Criteria for Special Inspection Agencies*, and are given as examples of qualifications. IAS AC291 criterion undergoes periodic revisions to keep pace with code changes, industry standards and changes in inspection protocol. The most current revision of IAS AC291 is available on the IAS website at www.iasonline.org/PDF/AC/AC291.pdf (see Table 1). Experience is hard to replace with education, and where the responsible professional for an agency provides his or her signature as evidence of competency for the special inspector, it should be respected. Ultimately, the IBC places the responsibility for approval of special inspectors and special inspection agencies upon the building official. In many jurisdictions, the obtaining of the IAS certificate of accreditation is sufficient for proof of competency.

A. Experience

1. In order for experience to count toward qualifications, it must be based on verifiable work directly related to the category or type of inspection involved.

2. An engineering degree (BS), in addition to appropriate in-house training, may be substituted for not more than one year of experience. An engineering technology degree (AA), in addition to appropriate in-house training may be substituted for not more than six months of experience. (Degree experience may not be substituted for more than half of the experience requirements in any category.)

3. Five or more years experience as a qualified special inspector in one or more categories of work may fulfill up to half of the experience requirements in any category, at the discretion of the (agency's) responsible professional engineer.

B. Certification

Certification, when specified, is intended to mean successful completion of an ICC examination appropriate to the category of work involved.

C. Special Inspector in Training

1. The intent of this provision is to provide practical opportunities for an inspector to gain the needed experience to qualify as a special inspector.

2. An inspector who does not meet the qualifications for special inspector may be allowed to perform "special inspection" at the discretion of the agency's responsible professional engineer, provided one or more of the following conditions are met:

   - Individual is working under the direct and continuous supervision of a special inspector fully qualified for the type of work involved.
   - Individual is working under the indirect or periodic supervision of a special inspector; and the scope of work is minor and/or routine and within the capabilities of the individual.
   - Individual is specifically approved by the building official.

TABLE 1
MINIMUM QUALIFICATIONS FOR SPECIAL INSPECTORS[a, b]

SPECIAL INSPECTION CATEGORY	REQUIRED EXPERIENCE	REQUIRED CERTIFICATION(S)	NOTES
Concrete construction (prestressed/precast)	Note c	ICC Prestressed SI and ICC Reinforced Concrete SI	
Reinforced concrete	Note c	ICC Reinforced Concrete SI or ACI Concrete Construction SI	
NDT	120 hours for Level II	ANSI/ASNT-CP-189 NDT or SNT-TC-1a NDT	
Pier and pile foundations	Note c	NICET II (geotechnical or construction, or construction material testing or soils)	Note d
Post-installed structural anchors in concrete	Note c	ICC Reinforced Concrete SI or ACI Concrete Construction SI	
Soils	Note c	ICC Soils SI or NICET II (geotechnical or construction, or construction material testing or soils)	Note d
Spray-applied, fire-resistant materials Intumescent, fire-resistant coatings Mastic, fire-resistant coatings	Note c	ICC Spray-applied Fireproofing SI or ICC Fire Inspector	
Steel (high-strength bolting)	Note c	ICC Structural Steel and Bolting SI	
Steel (welding)	5 years minimum or in accordance with AWS	AWS CWI or ICC Structural Welding SI	
Masonry construction	Note c	ICC Structural Masonry SI	
Wood construction	Note c	ICC Commercial Building Inspector or ICC Residential Building Inspector	Note d
EIFS	Note c	AWCI EIFS Inspector	
Firestop systems	Note c	UL firestop examination or FM firestop examination	
Wall panels, curtain walls and veneers	Note c	ICC Commercial Building Inspector or ICC Residential Building Inspector	Note d
Smoke control systems	Note d	AABC technician certification	Note d
Mechanical systems	Note c	ICC Commercial Mechanical Inspector or ICC Residential Mechanical Inspector	
Fuel-oil storage and piping systems	Note c	ICC Commercial Mechanical Inspector, ICC Residential Mechanical Inspector or API Above-ground Storage Tank Inspector	
Structural cold-formed steel	Note c	ICC Commercial Building Inspector or ICC Residential Building Inspector	Note d

(continued)

TABLE 1—continued
MINIMUM QUALIFICATIONS FOR SPECIAL INSPECTORS[a, b]

SPECIAL INSPECTION CATEGORY	REQUIRED EXPERIENCE	REQUIRED CERTIFICATION(S)	NOTES
Excavation—sheeting, shoring and bracing	Note c	NICET II (geotechnical or construction, or construction material testing or soils)	Note d
High-pressure steam piping (welding)	5 years minimum or in accordance with AWS	AWS CWI or ICC Structural Welding SI	
Structural safety—stability and mechanical demolition	Note c	RDP, PE, BS engineering/ architecture or valid site safety manager certification	
Site storm drainage disposal and detention	Note c	ICC Soils SI or NICET II (geotechnical or construction, or construction material testing or soils)	Note d
Sprinkler systems	Note c	ICC Commercial Building Inspector or ICC Residential Building inspector	Note d
Standpipe systems	Note c	ICC Commercial Building Inspector or ICC Residential Building Inspector	Note d
Heating systems	Note c	ICC Commercial Mechanical Inspector or ICC Residential Mechanical Inspector	Note d
Chimneys	Note c	ICC Commercial Mechanical Inspector or ICC Residential Mechanical Inspector	Note d
Seismic isolation systems	Note c	RDP, PE, BS engineering/ architecture	
Special cases	Note c	ICC Commercial Building Inspector or ICC Residential Building Inspector	Note d

a. It is recognized that the development of qualified inspectors requires those individuals to obtain experience performing inspections of actual work. The requirements herein include such experience, as do some of the required certifications. To provide a vehicle for individuals to obtain this experience, they shall perform inspections in accordance with written associate or apprentice programs that are prepared by the SIA, approved by the IAS and meet the requirements of the local governing authority. These programs must include, at a minimum, passing certification exams, when available, administered by third-party agencies, such as ICC and ACI; in-house SIA and third-party training; observation by the associate or apprentice of inspections performed by certified inspectors; and performance by the associate or apprentice inspectors of duplicate inspections with certified inspectors. This written program will also define the use of associate or apprentice inspectors and will limit their use based upon the level of supervision and the complexity of the inspection assignment. The complexity of an assignment shall be minimal and will often be task specific. Supervision should be direct, with a certified inspector being present at the site with the associate or apprentice. The associate or apprentice to certified inspector ratio on a project site shall not exceed 1:1. All documents related to work by an associate or apprentice inspector must be cosigned by a certified inspector. The written program must include documentation of compliance with the program.

b. When qualifications for special inspectors are locally defined, by statute, ordinance or rule that meet or exceed the requirements outlined in this criteria, these local requirements shall be recognized.

(continued)

TABLE 1—continued
MINIMUM QUALIFICATIONS FOR SPECIAL INSPECTORS[a, b]

c. Applicants shall comply with one of the following education and experience requirements:
   1. PE, licensed architects or RDP, and a minimum of three months of relevant work experience; or
   2. BS in engineering, architecture or physical science, and a minimum of six months of relevant work experience; or
   3. Two years of verified college or technical school (a copy of a diploma or transcript required), and a minimum of one year of relevant work experience; or
   4. High school or equivalent graduate (a copy of a diploma or certificate required), and a minimum of two years of verified relevant work experience; or
   5. A minimum of three years of verified relevant work experience.

d. PE, licensed architects or RDP are exempt from the required certifications listed in this table, but are subject to on-site assessment of competence by IAS.

Reference Abbreviations, and Recognized Certifying or Accreditation Agencies

AA	Associate of Arts (degree)
AABC	Associated Air Balance Council
ACI	American Concrete Institute
ANSI	American National Standards Institute
API	American Petroleum Institute
ASNT	American Society for Nondestructive Testing
ASTM	American Society for Testing and Materials
AWCI	Association of the Wall and Ceiling Industry
AWS	American Welding Society
BS	Bachelor of Science (degree)
CWI	Certified welding inspector
EIFS	Exterior insulation and finish system
FM	Factory Mutual Global
IAS	International Accreditation Service
IBC	*International Building Code*
ICC	International Code Council
ICC-ES	ICC Evaluation Service
NDT	Nondestructive testing
NICET	National Institute for Certification of Engineering Technologists
PE	Professional engineer
RDP	Registered design professional
SI	Special inspector
SIA	Special inspection agency
UL	Underwriters Laboratories

IAS
Accreditation Suite for Building Departments
FREE 6 PROGRAMS TO HELP BUILDING DEPARTMENTS APPROVE:
1. Special Inspection Agencies
2. Fabricators
3. Metal Building Manufacturers
4. Testing Laboratories
5. Field Evaluation Bodies
6. Building Department Service Providers
www.iasonline.org/suite
Toll Free 866-427-4422
ICC INTERNATIONAL CODE COUNCIL
IAS is a subsidiary of ICC
11-04418